HARIBANA

Harijanto
Setiawan

HARIBANA

Harijanto Setiawan

stichting kunstboek

Going bananas Xanthorrhoea johnsonii, Clematis, Musa

INTRODUCTION

Haribana was created in May 2020 in the mid of the global pandemic when the supply of fresh flowers and other materials was limited in Singapore. Used to making flamboyant designs with an abundance of flowers, the material shortage forced Harijanto to drastically rethink his design style and to come up with a fresh way to create high-quality and impactful designs using the least number of materials as possible. One could argue that the pandemic was a blessing in disguise for Harijanto, as it rekindled his passion for floristry and allowed him to reinvent himself and show another aspect of his personality as a designer – away from the extravagance everyone was used to. It allowed him to go back to his roots as an architect. While Harijanto's imagination was still limitless, the realization of his ideas had to be within the reality of the pandemic.
A grass leaf and a single flower are enough to create a Haribana piece, a highly architectural design statement with a single flower as focal point. The cherry on top, so to speak.

The name Haribana hints at Ikebana – even though these designs cannot officially be classified as 'Ikebana', as even the most free-spirited Ikebana schools objected to this. Haribana therefore is Harijanto's own interpretation of 'some sort of Ikebana' married with an architectural style and approach. Haribana pieces incorporate architectural concepts such as structure, balance, shape, line, color, texture, rhythm and harmony into floral art, as if the two are destined for each other, rather than forced upon.

Haribana techniques are various. By balancing, gluing, bending and folding foliage and grasses Harijanto first defines the lines and shapes of the piece. Steel grass stalks can be seen as the load bearing structures in architecture: the steel reinforcements and the concrete. The greenery is shaped to create the base structure, a filigree facade against which a single flower can shine. The addition of a stalk of flowers, handpicked by the artist, is like an architect adding his signature to a building.

Haribana isn't specifically aimed at a certain type of designer. It wants to be inclusive and universal. The style is a marriage of different cultures: it is a combination of the Oriental minimalist Zen look that can also be seen in traditional Ikebana, and European floral design techniques.
Unlike traditional flower arrangements, Haribana pieces aren't created for a specific spot in an interior. The shape of a piece is solely defined by the material that is available, but in the end it should still harmonize with the interior it is intended for.

Haribana is suitable for any designer who wants to explore and understand how architecture can be translated through flowers. Haribana has been well received and it is Harijanto Setiawan's hope that it will fill a unique void in floristry and develop into a specific style that is adapted worldwide.

The Creation of Adam *Xanthorrhoea johnsonii, Acer*

The Star Polygon *Xanthorrhoea johnsonii, Clematis*

The Gramophone *Xanthorrhoea johnsonii, Clematis*

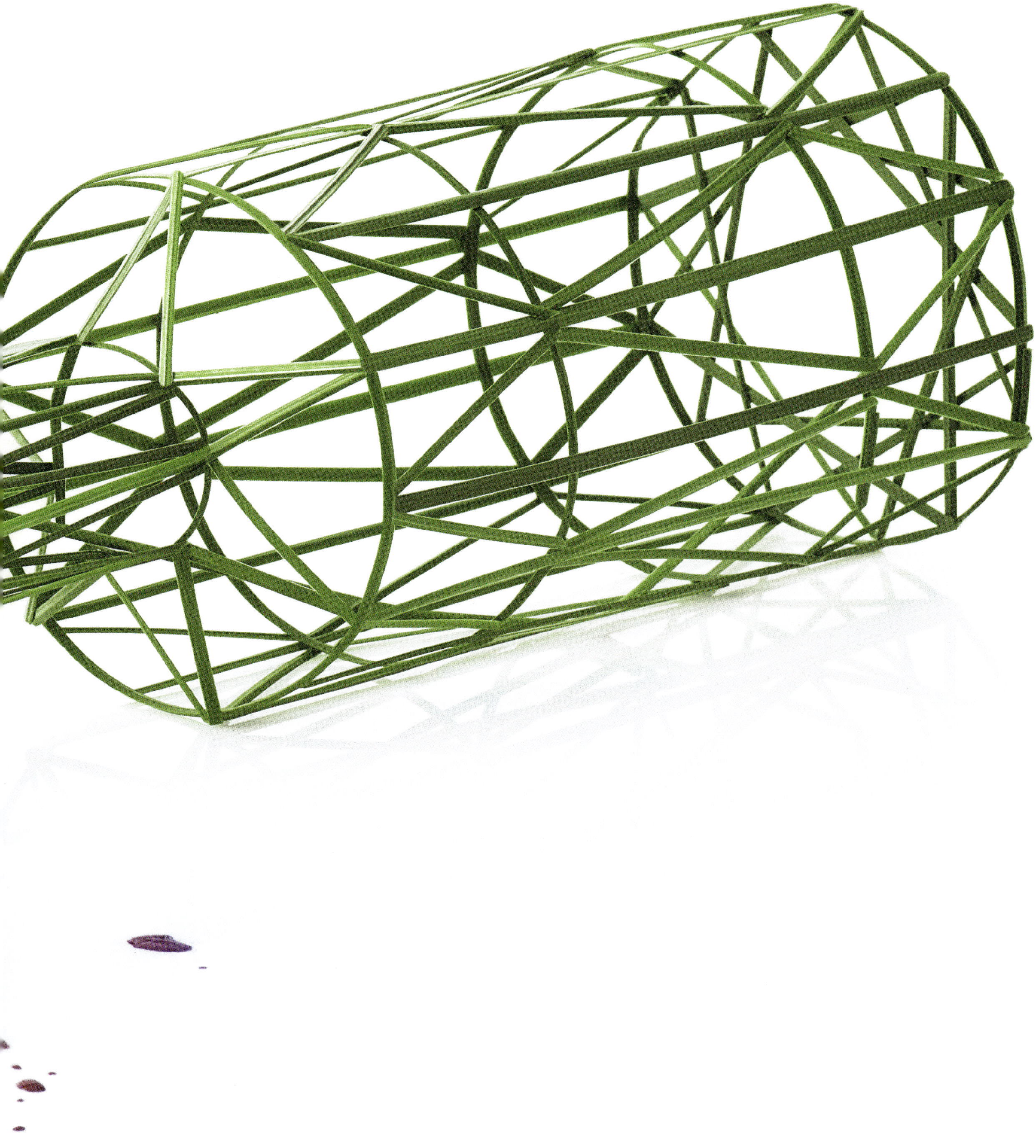

Bottleful Xanthorrhoea johnsonii, Clematis

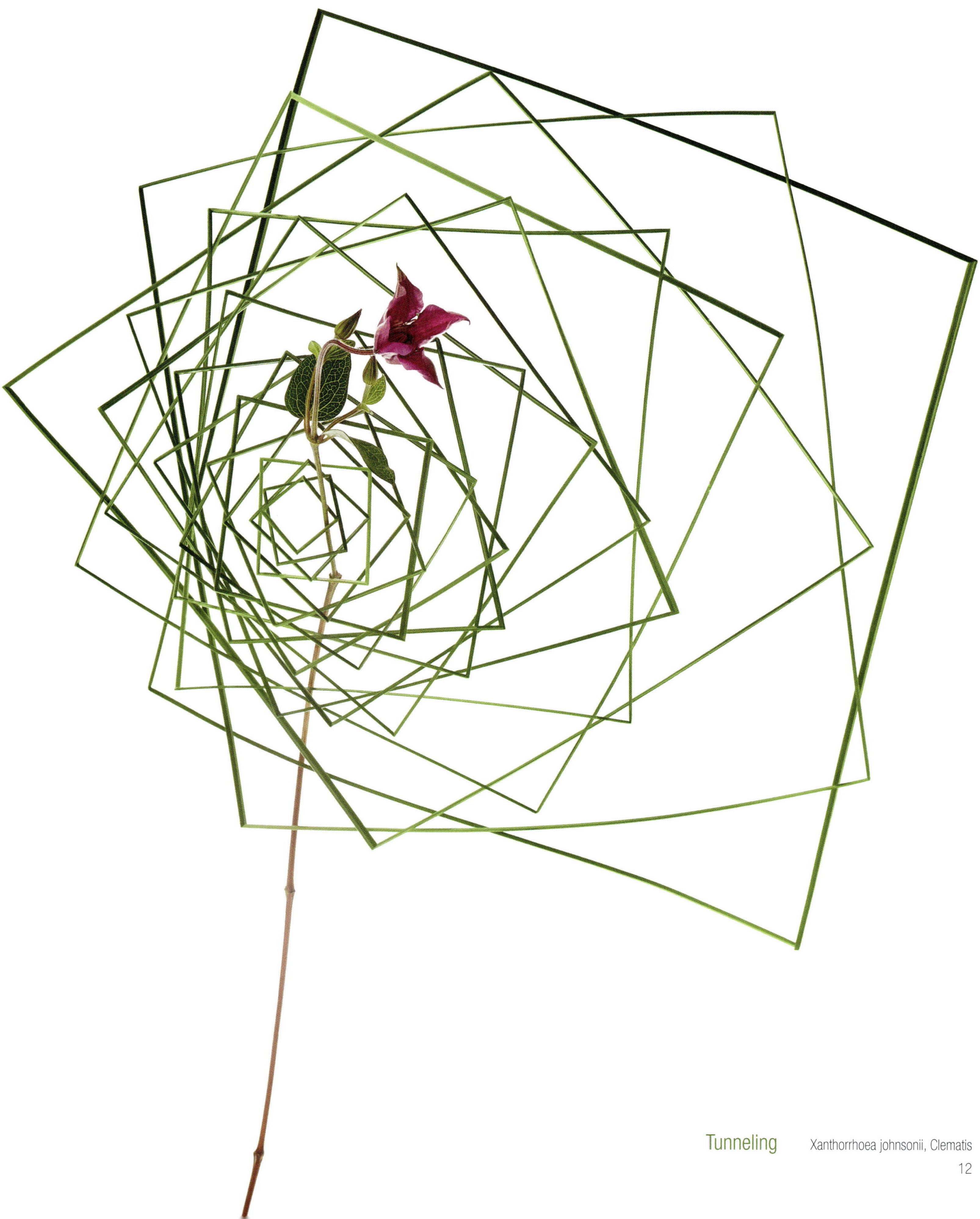

Tunneling *Xanthorrhoea johnsonii, Clematis*

Apex *Xanthorrhoea johnsonii, Allium neapolitanum*

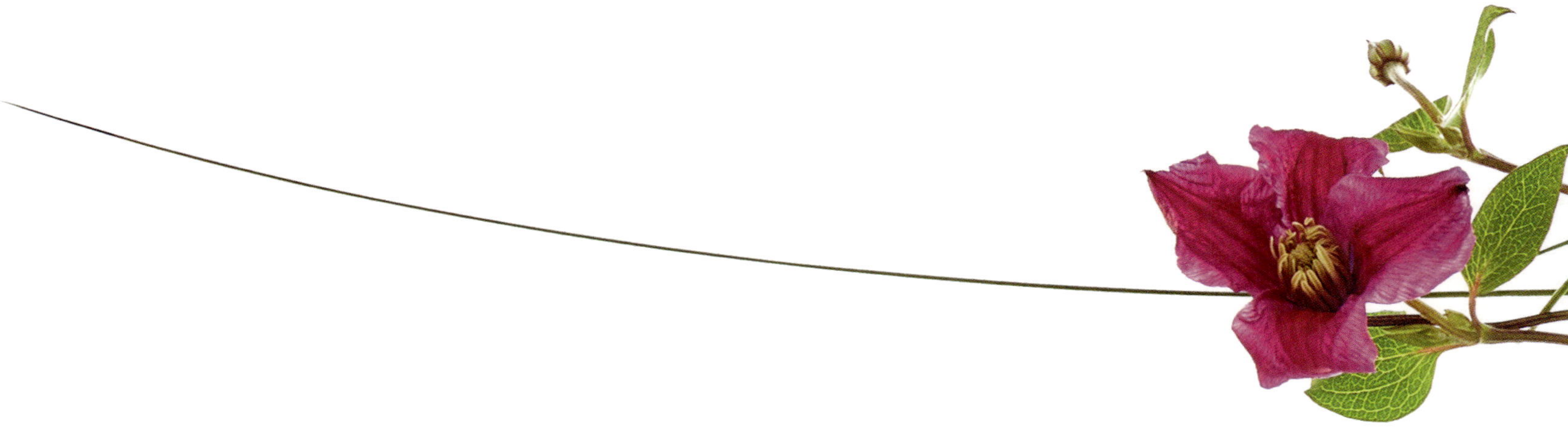

The Blowing Wind Xanthorrhoea johnsonii, Clematis

Dancing broom Xanthorrhoea johnsonii, Clematis

Fruition Clematis, Ficus benjamina

Cityscape *Xanthorrhoea johnsonii, Helleborus*

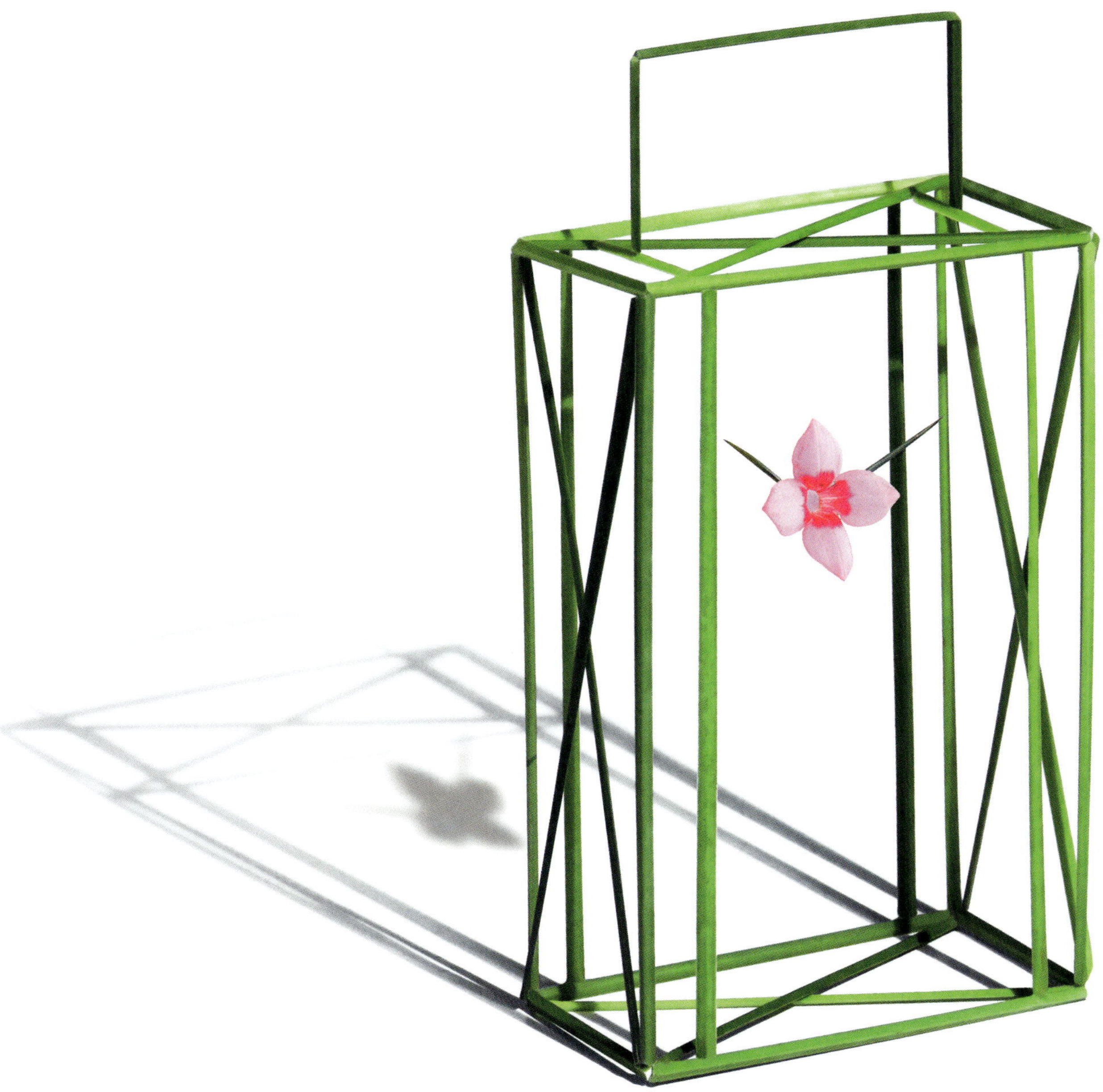

Table clock   Xanthorrhoea johnsonii, Bouvardia

Losing Gravity *Xanthorrhoea johnsonii, Clematis*

Sprouting Xanthorrhoea johnsonii, Clematis

Comet 2 Clematis, Pandanus baptistii

Two to Tango Crocosmia aurea, Clematis

Leap Xanthorrhoea johnsonii, Helleborus

In and Out *Xanthorrhoea johnsonii, Baldaran bulbosum, Helleborus orientalis*

Ellipse or Oval? Xanthorrhoea johnsonii, Iris germanica

Inverted Catenary Xanthorrhoea johnsonii, Clematis

Double Arch *Xanthorrhoea johnsonii, Clematis, Iris germanica*

Turbulence Xanthorrhoea johnsonii, Sandersonia aurantiaca, Fritillaria persica

Through Talinum

Stitching Nature Xanthorrhoea johnsonii, Clematis

Stabilized *Xanthorrhoea johnsonii, Anemone, Dracaena fragrans*

Gastronomy *Xanthorrhoea johnsonii, Nigella damascena*, Magnolia twigs

Sawing in half Clematis

Reflections Xanthorrhoea johnsonii, Clematis

Genie Xanthorrhoea johnsonii, Clemat s

Curiosity Pandanus baptistii, Allium sphaerocephalon, Stipa gigantea

Pointed Arch *Xanthorrhoea johnsonii, Monarda fistulosa*

Locomotion *Xanthorrhoea johnsonii, Clematis*

Brimming Xanthorrhoea johnsonii, Clematis

Variable Point Xanthorrhoea johnsonii, Monarda fistulosa, Equisetum hyemale, Clematis

Champagne Toast Xanthorrhoea johnsonii, Panicum virgatum, Allium neapolitanum

Rising Cuts *Monarda fistulosa, Equisetum hyemale, Aspidistra elatior*

Haven Aspidistra elatior, Clematis

Ribbon Rhythm Clematis, Pandanus baptistii

Ladder Xanthorrhoea johnsonii, Monarda fistulosa, Equisetum hyemale

Metro Line Xanthorrhoea johnsonii, Clematis

In Between Xanthorrhoea johnsonii, Iris germanica, Scabiosa atropurpurea

Outshine *Xanthorrhoea johnsonii, Clematis, Iris germanica*

Knock over   *Xanthorrhoea johnsonii, Centaurea cyanus*

Diagonal Matrix *Xanthorrhoea johnsonii, Delphinium elatum*

This And.... Xanthorrhoea johnsonii, Baldaran bulbosum, Delphinium

That   Xanthorrhoea johnsonii, Baldaran bulbosum, Asclepias

U Clematis, Dracaena fragrans, Heliconia

Comet *Xanthorrhoea johnsonii, Helleborus, Cyperus alternifolius*

Sea Creatures Clematis, Cordyline fruticosa

Out of the Cube Xanthorrhoea johnsonii, Scabiosa

Parallel Cyperus alternifolius, Bambusoideae, Magnolia grandiflora

Mind Blowing   Gladiolus acidanthera, Tulipa

Left or Right Clematis, Dracaena fragrans

Flutes Clematis, Baldaran bulbosum

Dancing Queen *Xanthorrhoea johnsonii, Sandersonia aurantiaca*

Luna *Xanthorrhoea johnsonii, Salix purpurea*

Swoosh Aspidistra elatior, Monarda fistulosa

Spindly Clematis, Bambusoideae, *Cyperus alternifolius*, *Gladiolus acidanthera*

Diving Xanthorrhoea johnsonii, Phalaenopsis, Philodendron

Strength *Xanthorrhoea johnsonii, Clematis, Baldaran bulbosum*

Air Xanthorrhoea johnsonii, Clematis

Saint Halo *Xanthorrhoea johnsonii, Clematis, Heliconia*

Steep and Shallow Xanthorrhoea johnsonii Cyperus alternifolius, Allium sphaerocephalon, Heliconia, Aquilegia vulgaris

Crescent Xanthorrhoea johnsonii, Clematis, Iris germanica, Ficus benjamina

Wand *Xanthorrhoea johnsonii, Campanula, Ficus benjamina*

Notes Xanthorrhoea johnsonii, Clematis, Iris germanica

Six Gladiolus acidanthera, Anemone

Conversely *Xanthorrhoea johnsonii, Clematis, Pandanus baptistii, Anemone coronaria*

HARIJANTO SETIAWAN

Born in Indonesia. An architect by training. Achieved his Master of Real Estate in 1997 from the University of New South Wales (Australia).

Discovered his true passion in floristry. Founder of Floral Design Studio Boenga in Singapore. Acclaimed for his artistic vision in the area of the 'floral-scape'.

Harijanto's talents have been on show at numerous public and private events worldwide. His spectacular floral creations are regularly seen at celebrity and luxury weddings. Breathtakingly fluent in the language of flowers, Harijanto's works inspire as much as they entertain.
Harijanto is the first floral designer that was awarded Designer of the Year 2013 at the Singapore Istana by the President of the Republic of Singapore. He continues to develop as floral demonstrator; he is an international judge for floral competitions and floral advisor for many international brands, design communities, institutions, luxury retail and hotels.

Harijanto's first book *Mystical Floral Journey*, published in 2016, documented his large-scale projects: from winning competition pieces to enormous wedding set-ups. In 2018 he published a second book *WTF! What The Flowers* that had a refreshing, light-hearted cocktail of floral arrangements. This third book *Haribana* highlights another aspect of Harijanto's floral design skills and personality.

HIGHLIGHTED ACCOLADES

Silver Leaf Award in the *International Floral Art 2010/2011*	Stichting Kunstboek, Belgium
Best Show and Gold Award	Singapore Garden Festival 2012
Best florist 2009/10/11/12/13/14/15/16/17/18	Singapore Tatler Magazine
International Designer of the Year 2015	Fusion Flowers, London
Silver Gilt Medal at RHS Chelsea Flower Show	London 2013
Designer of the Year 2013 - President's Design Award	Singapore
Gold Medal as Floral Ambassador at Royal Floria Putrajaya	Malaysia Flower Festival 2015
Gold Medal at Sanya Orchids Flower Show 2015	China
Best Colors Award At Goyang Flower Show 2017	Seoul, Korea